The Secrets of Mental Math

« *Everyone can do it in a flash!* »

Pascal IMBERT

Copyright © 2014, Pascal Imbert

All rights reserved

CreateSpace, ISBN-13: 978-1500662431

TABLE OF CONTENTS

INTRODUCTION 5

ADDITION TECHNIQUES 9

MAKE THE DISTINCTION BETWEEN A NUMBER AND A DIGIT...9
THE KEY: ADD FROM LEFT TO RIGHT 10
ADD FASTER BY THE REGROUPING METHOD 11
DECOMPOSE AN ADDITION WITH CLOSE NUMBERS 14
CUT OUT NUMBERS TO SOLVE AN ADDITION: GUARANTEED PERFORMANCE! 17
LEARN TO DETERMINE A DIGITAL ROOT 20
VERIFY AN ADDITION RESULT 22

SUBTRACTION TECHNIQUES....................25

SUBTRACT FROM LEFT TO RIGHT; IT IS SIMPLER 25
DECOMPOSE A SUBTRACTION WITH THE CLOSED NUMBERS 29
CUT OUT NUMBERS IN ORDER TO SOLVE A SUBTRACTION; IT IS SIMPLER 32
SUBTRACT NUMBERS TO 10, 100, 1000 IN THE TWINKLING OF AN EYE 35

DETERMINE THE CHANGE IN FRONT OF THE CASH REGISTER 39
VERIFY A SUBTRACTION RESULT 44

MULTIPLICATION TECHNIQUES 47

MULTIPLICATION TABLES - ALL YOU NEED 47
MULTIPLY BY 2, 4 AND 8 BEFORE THE CALCULATOR 49
MULTIPLY BY 5, 25 AND 50 BEFORE THE CALCULATOR 51
MULTIPLY BY 11 FASTER THAN ANYONE 53
MULTIPLY FROM LEFT TO RIGHT 57
MULTIPLY TO 3 DIGITS BY CLOSE NUMBERS 60
MULTIPLY BY DECOMPOSING: SIMPLIFY CALCULATION 65
SIMPLIFY DIGITS TO SOLVE A MULTIPLICATION 69
CUT OUT NUMBERS IN ORDER TO SOLVE A MULTIPLICATION 71
MENTAL MULTIPLICATION OF VERY LARGE NUMBERS: IT IS POSSIBLE 74
IMMEDIATELY ESTIMATE A MULTIPLICATION RESULT 81
VERIFY A MULTIPLICATION RESULT 83
SUM UP THE MULTIPLICATION TECHNIQUES 86

DIVISION TECHNIQUES 95

THIS NUMBER CAN BE DIVIDED? DIVISIBILITY
CRITERIA 96
DIVIDE BY 2, 4, AND 8 BEFORE THE
CALCULATOR 102
DIVIDING BY 5, 25, AND 50 BEFORE THE
CALCULATOR 104
DIVIDE BY 9 BEFORE THE CALCULATOR 107
DECOMPOSE NUMBERS IN ORDER TO SOLVE A
DIVISION 111
DIVIDE BY MEANS OF CLOSE NUMBERS 114
VERIFY A DIVISION RESULT 118

CONCLUSION **121**

Introduction

Mathematics, for many of us, is an unpleasant memory from school days. Many people, when evoking the word mathematics, have cold sweats, and they see themselves faced to a blackboard or an exam scratching their heads and asking themselves how to understand and solve the given problem. For all that, mathematics is not restricted to school, and it invades our daily lives, even if that is not obvious at first. Many moments of our daily life brings us to use numbers. For example, a classic day:

Early in the morning, you go to your favorite newsagent in order to buy your daily newspaper. You pay with a 5 Euros banknote; could you easily determine the change he is going to give you?

Then, you go to the supermarket. You fill your basket with ten products. Are you capable of estimating the approximate price that will be on your receipt?

Thereafter, you go to your post office because you need eleven 0.63-cent stamps. How many coins do you need to prepare to pay for your purchase?

Back home, you decide to make lunch and look for inspiration in a recipe book. The recipe you choose lists the ingredients needed for 6 people. As for you, you cook only for 4 people. Do you know how to easily determine which quantity of each ingredient you should use?

In the evening, you decide to plant a lawn in your garden. How would you easily know how many seed bags you should buy?

All of those examples have a common point: they show you the importance of mental arithmetic. In any of those situations, it is not necessarily practical to use a calculator.

Moreover, it is scientifically proven that appealing to your brain permits you to keep it trained and to increase its capacities. Consequently, I commit to bring you back together with mathematics and mental arithmetic, whatever your current level!

Do you know that most of the operations, whether it be additions, subtractions, multiplications, or divisions, can be mentally solved even if you abandoned mathematics a long time ago or didn't ever understand it? The only condition you need to be able to do it is to know some techniques allowing you to considerably simplify the given problem and to have some training.

In this book, I'm going to teach you these techniques. With a little practice and whatever your current skills in math are, you will be able to solve mentally, and very quickly, any calculation, even on 3-digit numbers. After reading this book, you will change the way you look at numbers, and some current complex situations will seem clearer to you. You will surprise your family and your friends by mentally solving some calculations in only a few seconds.

This book is written according to the reader's development, which is progressive. Every notion mentioned in a chapter leans on elements already mentioned in previous chapters. Take the time then, to read this book chapter after chapter, to well understand the techniques explained there and to solve the different exercises. Do not go quickly through from one chapter to another, and do not cut corners. By observing this advice, you will quickly progress, and you will be surprised by your performance and mental calculating ability.

Addition Techniques

Make the distinction between a number and a digit

Later in this book, we are going to talk about digits and numbers. That is why, in order for you to follow me, it is important to clarify the difference between these two words:

A **digit** goes from 0 to 9: so there are 10 digits, which are 0, 1, 2, 3, 4, 5, 6, 7, 8 and 9.

A **number** consists of a group of end-to-end digits: for example, 1256 is a 4-digit number, composed of the digits 1, 2, 5 and 6. Same way 658954 is a 6-digit number, composed of the digits 6, 5, 8, 9, 5 and 4.

If that is clear, we can move forward to the next stage.

The key: add from left to right

Most of the calculation techniques taught at school are adapted to written problem solving. In this way, it is common to do operations from right to left. For all that, daily life situations are more suitable for mental arithmetic.

Yet, one of the first pieces of advice we can give to anyone wishing to develop his mental arithmetic capacities is getting into the habit of doing calculations from left to right.

In writing, it is possible to add 4500 + 67 from right to left; however, when the calculation is done mentally, it is more natural and faster doing it from left to right.

This advice applies whatever the operation type you wish to do: addition, subtraction, multiplication, and division.

In mental arithmetic, always solve operations from left to right.

Add faster by the regrouping method

You would more easily solve additions if you could identify digits that, added together, give 10. Digits to associate are:

1	9	1 + 9 = 10
2	8	2 + 8 = 10
3	7	3 + 7 = 10
4	6	4 + 6 = 10
5	5	5 + 5 = 10

Figure 1 : Regrouping table

So, when solving an addition, begin by mentally grouping numbers that end in complementary digits.

Illustration:

Your shopping list includes 6 articles, as pointed below:

Toothpaste	*7 €*
Shower gel	*6 €*
Meats	*31 €*

Vegetables	24 €
Drinks	23 €
Frozen food	19 €

Solving the addition of this cashier ticket, group articles according to its price's last digit, as pointed out in the regrouping table (figure 1), which result in:

Toothpaste	7 €
Drinks	23 €
Shower gel	6 €
Vegetables	24 €
Meats	31 €
Frozen food	19 €

In this way, it is easier to solve additions 2 by 2 as:
toothpaste + drinks = 7 + 23 = 30 €
shower gel + vegetables = 6 + 24 = 30 €
meats + frozen food = 31 + 19 = 50 €

That is to say a total of 110 €.

Exercises:

Do the regroups, allowing you to facilitate the calculation of:

a/ 46 + 25 + 53 + 4 + 37 + 15
b/ 12 + 21 + 14 + 39 + 16 + 28
c/ 105 + 33 + 60 + 10 + 25 + 47

Answers:
a/ (46 + 4) + (25 + 15) + (53 + 37) = 50 + 40 + 90 = 180
b/ (12 + 28) + (21 + 39) + (14 + 16) = 40 + 60 + 30 = 130
c/ (105 + 25) + (33 + 47) + (60 + 10) = 130 + 80 + 70 = 280

Decompose an addition with close numbers

The regrouping method shows that it is easier to solve additions as soon as we see numbers containing 0 in the calculation.

In this way, when some addition numbers are close to whole numbers such as 100, 200, 300... 1000 etc..., it is very useful to handle these numbers in order to considerably simplify the calculation solving.

The game objective here is to modify the addition in order to write it in a more legible and more explicable way for our brain.

Illustration:

Suppose that we have this addition to solve: 356 + 696.

Mentally, you identify that 696 is close to 700:
696 = 700 - 4

The proposed addition can then be written in a simpler way, that is:

356 + 700 - 4

Calculating from left to right: 1056 − 4 = 1052

**Use the same principle to calculate:
204 + 387 + 615**

Mentally, you identify that:
204 is close to 200 (204 = 200 + 4)
387 is close to 400 (387 = 400 − 13)
615 is close to 600 (615 = 600 + 15)

We can then rewrite the addition this way:
200 + 4 + 400 − 13 + 600 + 15

Using the regroup method: 200 + 400 + 600 + 4 − 13 + 15

That is to say 1200 + 6 = 1206

Exercises:

Solve these next additions by the decomposition method:

a/ 196 + 742
b/ 203 + 536 + 488

Answers:
a/ 200 − 4 + 742 = 942 − 4 = 938
b/ 200 + 3 + 500 + 36 + 500 − 12 = 1200 + 27 = 1227

Use the decomposition method by close numbers to simplify additions making appear 0.

Cut out numbers to solve an addition: guaranteed performance!

A little known but very efficient method to simplify a difficult addition into 2 or 3 simpler additions consists of cut out numbers.

When mentally calculating, cutting out numbers can considerably simplify the calculation's difficulty.

Illustration:

Suppose you have this addition to solve: 326521 + 432478.

Mentally, you undertake the next cutting out:
32 / 65 / 21
43 / 24 / 78

And you add numbers cut out this way:
32 + 43 = 75
65 + 24 = 89
21 + 78 = 99

Which results in putting the results end-to-end: 758999.

Sometimes, calculation is a little more complicated because there is a remainder as when you have to calculate 541268 + 39323

Mentally, you undertake the next cutting out:
54 / 12 / 68
03 / 93 / 23 *(we add a 0 at the beginning in order to make sure the 2 numbers to be added each have 6 digits)*

And you add numbers cut out this way:
54 + 03 = 57
12 + 93 = **1**05 *(you should keep only 2 digits, the **1** will add to the previous 57, what results in 58)*
68 + 23 = 91

What results in 580591

Exercises:

Solve next additions by the cutting out method:

a/ 541247 + 251478
b/ 32569 + 4781
c/ 365214 + 325874

Answers:
a/ 54/12/47 + 25/14/78 = 54 + 25 / 12 + 14 / 47 + 78 = 79 / 26 / **1**25 = 79 / 27 / 25 = 792725

b/ 3/25/69 + 47/81 = 3 / 25 + 47 / 69 + 81 = 3 / 72 / **1**50 = 3 / 73 / 50 = 37350
c/ 36/52/14 + 32/58/74 = 36 + 32 / 52 + 58 / 14 + 74 = 68 / **1**10 / 88 = 69 / 10 / 88 = 691088

Use the cutting out numbers method to simplify larger numbers.

Learn to determine a digital root

Leave out calculation methods to get interested in the digital root. This notion which name is, I agree, a little bit barbarian and very simple to learn and will serve you in your daily life.

You have already flipped through last pages of a women magazine and you have stopped at the Numerology section. In this section you learn about what the future offers to you, according to your lucky number between 0 and 9.

This number is very easy calculated from your birthday date. Suppose your birthday date is August, 20th 1976 or 08/20/1976. You add then all digits from this date: 2 + 0 + 0 + 8 + 1 + 9 + 7 + 6, that is to say 33. You add now digits from this number 3 + 3 = 6. If you were born on august 20th, 1976, your lucky digit is then number 6.

This digit is called the **digital root**. It is obtained by successive additions of digits composed of a number to the obtaining of a digit composed between 0 and 9.

Exercises:

Determine the digital root of these numbers:

a/ 54126
b/ 8745
c/ 236514

Answers:
a/ 5+4+1+2+6=18 → 1+8 = 9
b/ 8+7+4+5=24 → 2+4 = 6
c/ 2+3+6+5+1+4=21 → 2+1 = 3

The digital root is a digit between 0 and 9 as a result of the successive addition of summing digits.

Verify an addition result

In the previous chapter, I sold you the digital root as a tool very useful in daily life. In fact, you could use it after each calculation, whatever it is: additions, subtractions, multiplications, or divisions, in order to make sure that the result you get is correct.

I'm going to now show you how to use the digital root in order to verify an addition result.

Illustration:

Previously, we calculated 326521 + 432478 = 758999

Calculate the digital root and its result:
326521: 3+2+6+5+2+1=19 → 1+9=10 → 1+0 = 1
432478: 4+3+2+4+7+8=28 → 2+8=10 → 1+0 = 1
758999: 7+5+8+9+9+9=47 → 4+7=11 → 1+1 = 2

The operation we made (326521 + 432478) is an addition.

We then add the digital root of 326521 (which is 1) and that of 432478 (which is 1), that is to say 2.

We verify that the obtained digital root is equal to the digital root of the result 758999 (which is 2).

If we get the same digital root, it means that the calculation is probably correct; if it is not the case, it is certain that the calculation is wrong.

Exercises:

Determine, by the digital root method, if these calculations are wrong:

a/ 359 + 423 = 782
b/ 1026 + 478 = 1504
c/ 586 + 1234 = 1830

Answers:
a/
Determine the digital roots:
359: 3+5+9=17→1+7=8
423: 4+2+3=9
782: 7+8+2=17→1+7=8
Add the digital roots:
8+9=17→1+7=8
The digital roots are the same; the calculation is certainly correct.

b/
Determine the digital roots:
1026: 1+0+2+6=9
478: 4+7+8=19→1+9=1

1504: 1+5+0+4=10→1+0=1
Add the digital roots:
9+1=10→1+0=1
The digital roots are the same; the calculation is certainly correct.

c/
Determine the digital roots:
586: 5+8+6=19→1+9=10→1+0=1
1234: 1+2+3+4=10→1+0=1
1830: 1+8+3+0=12→1+2=3
Add the digital roots:
1+1=2
The digital roots are not the same; the calculation is certainly wrong.

Determine if an addition result is wrong by using the digital roots.

Subtraction techniques

Subtract from left to right; it is simpler

We have said that in mental arithmetic, it is easier to calculate from left to right.

There are many advantages to doing calculations from left to right because we pronounce and write numbers from left to right. However, sometimes, we only need the first important digits, and that would be a waste of time doing all those calculations, as it is the case when we begin from the right.

Illustration:

Suppose that you have this subtraction to solve:

$$\begin{array}{r} 62 \\ -\ 47 \end{array}$$

Mentally, you undertake the calculation from left to right:

6 − 4 = 2

Seeing that, in the next column 2 − 7 will not be possible, take away 1 to the obtained result so 2 − 1 = 1;

Now, in the second column, calculate 12 − 7 instead of 2 − 7: 12 − 7 = 5

Putting results end-to-end, this equals 15.

This subtraction method can be extended to larger numbers:

$$\begin{array}{r} 41268 \\ -\ 39323 \end{array}$$

Beginning from the left:

4 − 3 = 1, but as the next column calculation is not possible (1 − 9), we take away 1 from the result 1 − 1 = 0 and in the next column, we will calculate 11 − 9 instead of 1 − 9,

In the second column, we calculate 11 − 9 = 2 but as the next column calculation is not possible (2 − 3), we take away 1 from the result 2 − 1 = 1, and in the next column, we will calculate 12 – 3 instead of 2 - 3.

In column 3, we calculate 12 – 3 = 9, and we realize that the next column calculation is possible (6 – 2), so we keep this result (9).

In column 4, we calculate 6 – 2 = 4 and we realize that the next column calculation is possible (8 – 3), so we keep this result (4).

In column 5, we calculate 8 – 3 = 5

We then put all obtained digits end-to-end, getting 01945, which is 1945. We conclude that 41268 – 39323 = 1945.

Exercises:

Solve these subtractions from left to right:

a/ 2568 - 1243
b/ 7236 - 4412
c/ 5214 - 1875

Answers:
a/ 1325
b/ 2824
c/ 3339

Solve your subtractions from left to right:

Subtract each column, beginning from the left, but before writing the answer, look at the next column:
- If the upper is lower than the one in the bottom, write the answer.
- If not, you simplify the digit by 1, write the result, and give the other 1 to the smaller upper number of the next column.
- If the digits are the same, look at the next column to decide how to continue.

Decompose a subtraction with the closed numbers

As in an addition, a subtraction can be simplified from the moment we see numbers containing 0 in the calculation.

The game objective is to modify the subtraction in order to write it in a more legible and easier way to be interpreted by our brain. In fact, it has a natural capacity to see how much things lack in order to fill an empty space.

In this way, 98 is close to 100, but it lacks 2, just as 389 is close to 400 but lacks 11. This observation permits us to calculate in a simpler way.

Illustration:

Suppose that you have this subtraction to solve: 54 - 18.

Mentally, you identify that 18 is close to 20:
18 = 20 - 2

The proposed subtraction can then be written under a simpler form, which is:
54 - 20 + 2 (when taking away 20 instead of 18, we take away 2 more)

That is to say: 34 + 2 = 36

**Use the same principle to calculate:
967 - 401 - 198**

Mentally, you identify that:
401 is close to 400 (401 = 400 + 1)
198 is close to 200 (198 = 200 - 2)

We can then rewrite the addition under this form:
967 − (400 + 1) − (200 − 2) = 967 − 400 − 1 − 200 + 2

Using the regrouping method: 967 - 400 - 200 - 1 + 2

That is to say: 367 + 1 = 368

Exercises:

Solve the next subtractions by the decomposing method:

a/ 536 - 122
b/ 846 - 295 - 197

Answers:
a/ 536 − 100 - 22 = 436 − 22 = 414
b/ 846 - 300 + 5 - 200 + 3 = 346 + 8 = 354

Use the decomposing method by close numbers in order to simplify subtractions that contain 0.

Cut out numbers in order to solve a subtraction; it is simpler

As in an addition, the mental arithmetic in subtractions by cutting out numbers can considerably simplify the calculation difficulty.

Illustration:

Suppose you have this subtraction to solve:

$$541236$$
$$- 251012$$

Mentally, you can undertake the next cutting out:
541236: 54 / 12 / 36
251012: 25 / 10 / 12

And you subtract the cut out numbers:
54 - 25 = 29
12 - 10 = 02
36 - 12 = 24

Putting results end-to-end equals 290224.

Sometimes, calculation seems to be a little more complicated, as in the case when you have to calculate:

845219
- 632587

In fact, if you cut out the numbers 2 by 2, 84/52/19 and 63/25/87, the last calculation to solve will be 19 – 87 ...
Likewise, if you cut out the numbers 3 by 3, 845/219 and 632/587, the last calculation to solve will be 219 – 587 ...

The best cutting out is the next one:
84 / 521 / 9
63 / 258 / 7

And you subtract the cut out numbers:
84 - 63 = 21
521 - 258 = 521 – 200 – 58 = 321 – 58 = 263
9 - 7 = 2

That is to say: 212632.

It is then important to determine from the beginning which is the wiser cutting out.

Exercises:

Solve the next subtractions by a wise cutting out method:

a/ 378514 - 227309
b/ 44625 - 21563
c/ 87452 - 63247

Answers:
a/ 37/85/14 – 22/73/09 = 37 - 22 / 85 – 73 / 14 – 09 = 15 / 12 / 05 = 151205
b/ 44/62/5 – 21/56/3 = 44 - 21 / 62 - 56 / 5 – 3 = 23 / 06 / 2 = 23062
c/ 8/74/52 – 6/32/47 = 8 – 6 / 74 – 32 / 52 – 47 = 2 / 42 / 05 = 24205

Use the number cutting out method in order to simplify the large number subtractions.

Subtract numbers to 10, 100, 1000 in the twinkling of an eye

There is an infallible method to solve subtractions from numbers such as 10, 100, 1000, 10000, 100000.... You could challenge your friends, announcing to them that you are capable of giving them the result of 1000000 − 652147 before they can use their calculator. So, how to succeed at this feat?

Illustration:

We are then going to get interested in the next subtraction:

$$1000000$$
$$-\ 652147$$

This technique consists in looking at the bottom number (652147) and determining for each digit its complement to 9 and for the last digit its complement to 10:

What is a complement to 9?
It is the digit to add to another digit in order to get 9.

For example, the 9 complement of 7 is 2 (this is 7+2=9) while the 9 complement of 4 is 5 (this is 4+5=9).
What is a complement to 10?
In the same way, it is the digit to add to another digit in order to get 10.
For example, the 10 complement of 6 is 4 (this is 6+4=10) while the 10 complement of 2 is 8 (this is 2+8=10).

The number to be subtracted, in our example, is 652147
Complement of 6 to 9 → 3
Complement of 5 to 9 → 4
Complement of 2 to 9 → 7
Complement of 1 to 9 → 8
Complement of 4 to 9 → 5
Complement of 7 to 10→ 3

Putting digits end-to-end equals 347853

So 1000000 – 652147 = 347853.

Important Observations: the bottom number should always have as many digits as 0 in the upper number.

Calculate for example:

$$\begin{array}{r} 10000 \\ - \quad 89 \end{array}$$

The upper number has 4 zeros while the bottom number has 4 digits only. In this calculation, it is necessary to read 2 zeros to the bottom number as bellow:

$$\begin{array}{r} 10000 \\ -0089 \end{array}$$

The number to be subtracted, in our example, is 0089
Complement of 0 to 9 → 9
Complement of 0 to 9 → 9
Complement of 8 to 9 → 1
Complement of 9 to 10 → 1

Putting digits end-to-end, this equals 9911

That is to say: 10000 − 89 = 9911.

Exercises:

Solve the next subtractions by the based on 10 method:

a/ 100000 - 52478
b/ 10000 - 1563
c/ 100000 - 247

Answers:
a/ 100000 - 52478 = 47522
b/ 10000 − 1563 = 8437
c/ 100000 − 247 = 100000 − 00247 = 99753

When you have to subtract a number of 10, 100, 1000 ... (method based on 10):

Make sure that the bottom number has as many digits as the upper number has zeros,
If this is not the case, re add zeros to the bottom number,
Calculate the complement to 9 of the bottom number digits and the complement to 10 of the last digit of the bottom.
These digits put end-to-end are the subtraction result.

Determine the change in front of the cash register

There is an exercise you could apply to your daily life in order to stimulate your brain to solve mental arithmetic. It is to determine how much the change the storekeeper is going to give you will be when you do the shopping and you pay with a banknote.

Mastering this calculation technique will allow you to make sure that you receive the correct change. Do not forget that good accounts make good friends!

Illustration:

You give a 10€ banknote to pay a 6.53€ purchase.
In order to determine the given change, you solve this calculation:

$$10.00 - 6.53$$

Applying the based on 10 method studied earlier, we get: 3.47€

***

Now, you give a 20€ banknote to pay your 12.54€ purchase

To determine the given change, you solve this calculation:

$$20.00$$
$$-\ 12.54$$

Related to the based on 10 method, the variant consists of treating differently each number's first digit:

To the first digit of the upper number (2) we subtract the first digit of the bottom number (1) + 1 = 2, that is to say 0

Then, we apply the based on 10 method to the other digits of the bottom number (2, 5 and 4) situated under the upper number zeros, that is to say: 746.

Putting digits end-to-end, we get 07.46 so 7.46€

***

Now, you give a 50€ banknote to pay your 29.95€ purchase

To determine the given change, you solve this calculation:

$$\begin{array}{r} 50.00 \\ -\ 29.95 \end{array}$$

To the upper number's first digit (5) we subtract the bottom number's first digit (2) + 1 = 3, that is to say 2.

Then, you apply the based on 10 method to the other bottom digits (9, 9 and 5) situated under the upper number zeros, that is to say: 005.

Putting digits end-to-end, we get 20.05 so 20.05€.

***

Finally, you give a 200€ banknote to pay your 78.32€ purchase

To determine the given change, you solve this calculation:

$$\begin{array}{r} 200.00 \\ -\ 078.95 \end{array}$$

We realize that the bottom number has fewer digits (4) than the upper number (5). We add then a 0 to the bottom number.

To the upper number's first digit (2) we subtract the bottom number's first digit (0) + 1 = 1, that is to say: 1.

Then, we apply the based on 10 method to the other bottom digits (7, 8, 9 and 5) situated under the upper number zeros, that is to say: 2105.

Putting digits end-to-end, we get 121.05, so 121.05€.

Exercises:

Determine the given change in each case:

a/ 10 - 3,69
b/ 50 - 25,44
c/ 200 - 24,72

Answers:
a/ 6.31
b/ 24.56
c/ 175.28

To determine the given change, apply the based on 10 method or its variant to multiples of 10, for example, 20, 50 or 200.

Verify a subtraction result

As in an addition, it is possible to use the digital root in order to verify a subtraction result.

Illustration:

Previously, we calculated 845219 - 632587 = 212632

Calculate the digital roots of numbers below:
845219: 8+4+5+2+1+9=29 → 2+9=11
→ 1+1 = 2
632587: 6+3+2+5+8+7=31 → 3+1=4
212632: 2+1+2+6+3+2=16 → 1+6=7

The operation we did (845219 - 632587) is a subtraction.

*We subtract then the digital root of 845219 (which is 2) and the one of 632587 (which is 4), that is to say -2. If the result is negative, it is necessary to add to it a **9,** that is to say -2 + 9 = 7. If the result is positive, it could be used.*

We verify that the gotten digital root is the same as the result digital root 212632 (which is 7).

If we get the same digital root, calculation is probably correct; if it is not the case, calculation is certainly wrong.

Exercises:

Determine, by the digital method, if these calculations are wrong:

a/ 658 - 312 = 346
b/ 3627 - 1265 = 2372
c/ 4797 - 524 = 4273

Answers:
a/
Determine digital roots:
658: 6+5+8=19→1+9=10→1+0=1
312: 3+1+2=6
346: 3+4+6=13→1+3=4
Subtract digital roots:
1-6=-5→ a negative result, we add 9 → -5+9=4
The digital roots are the same; calculation is correct.

b/
Determine digital roots:
3627: 3+6+2+7=18→1+8=9
1265: 1+2+6+5=14→1+4=5
2372: 2+3+7+2=14→1+4=5
Subtract digital roots:
9-5=4
The digital roots are not the same; calculation is certainly wrong.

c/
Determine digital roots:
4797: 4+7+9+7=27→2+7=9
524: 5+2+4=11→1+1=2
4273: 4+2+7+3=16→1+6=7
Subtract the digital roots:
9-2=7

The digital roots are the same; calculation is correct.

Determine if a subtraction result is wrong by using digital roots.

Multiplication techniques

Multiplication tables - all you need

The first good news in this chapter is that you need only one tool to be able to calculate complex multiplication results. This tool is called "multiplication tables". It is the only indispensable knowledge without which you cannot expect to solve a multiplication. The rest is only techniques and tricks to be taught to you in the next pages.

First of all, I remind you the multiplication b.a.-ba. Take all the time you need, but make sure you know by heart the next tables:

2 x 1 = 2	3 x 1 = 3	4 x 1 = 4	5 x 1 = 5	6 x 1 = 6
2 x 2 = 4	3 x 2 = 6	4 x 2 = 8	5 x 2 = 10	6 x 2 = 12
2 x 3 = 6	3 x 3 = 9	4 x 3 = 12	5 x 3 = 15	6 x 3 = 18
2 x 4 = 8	3 x 4 = 12	4 x 4 = 16	5 x 4 = 20	6 x 4 = 24
2 x 5 = 10	3 x 5 = 15	4 x 5 = 20	5 x 5 = 25	6 x 5 = 30
2 x 6 = 12	3 x 6 = 18	4 x 6 = 24	5 x 6 = 30	6 x 6 = 36
2 x 7 = 14	3 x 7 = 21	4 x 7 = 28	5 x 7 = 35	6 x 7 = 42
2 x 8 = 16	3 x 8 = 24	4 x 8 = 32	5 x 8 = 40	6 x 8 = 48
2 x 9 = 18	3 x 9 = 27	4 x 9 = 36	5 x 9 = 45	6 x 9 = 54
2 x 10 = 20	3 x 10 = 30	4 x 10 = 40	5 x 10 = 50	6 x 10 = 60

7 x 1 = 7	8 x 1 = 8	9 x 1 = 9	10 x 1 = 10
7 x 2 = 14	8 x 2 = 16	9 x 2 = 18	10 x 2 = 20
7 x 3 = 21	8 x 3 = 24	9 x 3 = 27	10 x 3 = 30
7 x 4 = 28	8 x 4 = 32	9 x 4 = 36	10 x 4 = 40
7 x 5 = 35	8 x 5 = 40	9 x 5 = 45	10 x 5 = 50
7 x 6 = 42	8 x 6 = 48	9 x 6 = 54	10 x 6 = 60
7 x 7 = 49	8 x 7 = 56	9 x 7 = 63	10 x 7 = 70
7 x 8 = 56	8 x 8 = 64	9 x 8 = 72	10 x 8 = 80
7 x 9 = 63	8 x 9 = 72	9 x 9 = 81	10 x 9 = 90
7 x 10 = 70	8 x 10 = 80	9 x 10 = 90	10 x 10 = 100

Multiply by 2, 4 and 8 before the calculator

Multiplying by 2 is easy and can be used to quickly solve simple calculations. This way, 26 x 2 would be the same as 26 + 26 is 52.

Thanks to multiplication by 2, we can easily multiply by 4 and by 8. In fact, it is enough to note that:

Multiplying by 4 would be the same as multiplying two times by 2,
Multiplying by 8 would be the same as multiplying three times by 2.

In this way, 24 x 4 = 24 x 2 x 2 = 48 x 2 = 96

And 13 x 8 = 13 x 2 x 2 x 2 = 26 x 2 x 2 = 52 x 2 = 104

Exercises:

Solve these multiplications:

a/ 21 x 4
b/ 17 x 8
c/ 54 x 4

Answer:
a/ 84

b/ 136
c/ 216

**Multiplying by 2 is an easy to do operation.
Multiplying by 4 would be the same as multiplying two times by 2,
Multiplying by 8 would be the same as multiplying three times by 2.**

Multiply by 5, 25 and 50 before the calculator

During the study of addition techniques, we saw that there is a method that allows you to considerably simplify calculations and that consists in making zeros appear in the operation. In the case of a multiplication, it is easy to make zeros appear when we multiply 2 x 5, since the result is 10.

Keeping that in mind, it is very simple to solve multiplications by 5, by 50, and even by 25.

In fact:

In order to multiply by 5, begin with a multiplication by 10, and then divide by 2,
In order to multiply by 50, begin with a multiplication by 100, and then divide by 2,
In order to multiply by 25, begin with a multiplication by 100, and then divide two times by 2.

Illustration:

We need to calculate 18 x 5
Begin calculating 18 x 10 = 180
Then, we calculate 180 / 2 = 90
So 18 x 5 = 90

Calculate 6.2 x 50
Begin calculating 6.2 x 100 = 620
Then, we calculate 620 / 2 = 310

Calculate 246 x 25
Begin calculating 246 x 100 = 24600
Then, we calculate 24600 / 2 = 12300
And then again 12300 / 2 = 6150

Exercises:

Solve these multiplications:

a/ 22 x 5
b/ 16 x 50
c/ 1,4 x 25

Answers:
a/ 110
b/ 800
c/ 35

Multiplying by 5 would be the same as multiplying by 10 and then, divide by 2,
Multiplying by 50 would be the same as multiplying by 100 and then, divide by 2,
Multiplying by 25 would be the same as multiplying by 100 and then, divide two times by 2.

Multiply by 11 faster than anyone

There is a digit that you are going to love when solving multiplications. It is the number 11. Nowadays, many people tremble at the idea of solving, mentally, a multiplication with the digit 11. According to me, it is the type of situation I love. Soon, you are going to be able to challenge your friends by betting that you are capable of solving mentally a multiplication by 11 faster than them...

For the digits between 1 and 9, this is very simple, since the multiplication by 11 consists in doubling the digit we multiply, this way:

1 x 11 = 11 (we double 1)
2 x 11 = 22 (we double 2)
3 x 11 = 33 (we double 3)
And we continue like this up to 9 x 11 = 99

Nevertheless, what does it happen when multiplying by 11 less friendly digits as 0.63 or 321?

Illustration:

Today, you arrive to your office, and you want to buy some stamps. Each stamp has a face value of 0.63€, and you need 11 stamps. How much change should you prepare?

We are going to calculate the price to be paid for these stamps. A stamp costs 0.63€ so, 11 stamps cost 11 x 0.63:

We consider that 0.63€ equals to 63 cents. We calculate then 63 x 11.

In order to solve it, we begin framing 63 with zeros, that is to say:
063**0**

Next, starting from the right, we add digits 2 by 2, that is to say:

0 + 3 = 3
3 + 6 = 9
6 + 0 = 6

And we put the calculated digits end-to-end, which is 693. That is to say: 63 x 11 = 693.
In order to get the result of 0.63 x 11, we divide the previous result by 100, which is to say 6.93.

Therefore, 0.63€ 11 stamps will cost you 6.93€.

***

Calculate 58 x 11.

We mentally write 0580 and we add digits 2 by 2 starting from the right:

0 + 8 = 8
8 + 5 = **1**3, we keep the 3 and the **1** will be added to the next calculation.
5 + 0 = 5, to which we add the previous calculation **1**, which is 6.

We conclude then that 58 x 11 = 638.

***

That works with even bigger digits.

Calculate 6239 x 11

We mentally write 062390, and we add digits 2 by 2 starting from the right:

0 + 9 = 9
9 + 3 = **1**2, we keep the 2 and the **1** will be added to the next calculation.
3 + 2 = 5, to which we add the previous calculation **1**, which is 6.
2 + 6 = 8
6 + 0 = 6

We conclude then that 6239 x 11 = 68629.

Exercises:

Solve these multiplications:

a/ 22 x 11
b/ 685 x 11
c/ 56 x 11

Answers:
a/ 242
b/ 7535
c/ 616

**In order to multiply by 11,
Mentally frame your number with zeros.
Add digits 2 by 2 starting from the right.
Put digits end-to-end in order to form the result.**

Multiply from left to right

When solving a mental calculation, multiply from left to right rather than from right to left, as taught at school, to solve a written multiplication. This will allow you to simplify calculation by giving, from the beginning, a good estimation of the calculation result.

Illustration:

We need to calculate 241 x 4

Beginning from the left, we have:

2 x 4 = 8
4 x 4 = **1**6. We keep the 6, and the **1** is added to the previous calculation (the 8 becomes then 9)
1 x 4 = 4

From the first calculation, we know that the answer will be between 800 and 900. Then, we gradually refine the calculation in order to conclude that 241 x 4 = 964.

***

Calculate 745 x 3

Beginning from the left, we have:

7 x 3 = 21
4 x 3 = **1**2, we keep the 2 and the **1** is added to the previous calculation (21 becomes then 22).
5 x 3 = **1**5, we keep the 5 and the **1** is added to the previous calculation (2 becomes then 3).

We conclude that 745 x 3 = 2235.

Exercises:

Solve these multiplications from left to right:

a/ 431 x 3
b/ 124 x 6
c/ 12432 x 2

Answers:
a/ 1293
b/ 744
c/ 24864

Solving a multiplication from left to right, allow, from the beginning, estimating the result magnitude and it can simplify calculation.

Multiply to 3 digits by close numbers

What do these multiplications have in common: 24 x 22, 16 x 12, 56 x 52, 78 x 73, 91 x 95?

All these operations show the particularity to multiply two 2-digit numbers beginning with the same first digit: 24 and 22 begin with 2, 16 and 12 begin with 1, 56 and 52 begin with 5, 78 and 73 begin with 7, finally, 91 and 95 begin with 9.

If you can identify these situations, you will be in a great position in order to mentally solve these operations because there is a calculation technique of close number multiplication.

Illustration:

Calculate 24 x 22

We notice that these 2 numbers are close to 20, in fact:
24 = 20 + 4
22 = 20 + 2

*To 24 we add the **2** from 22 = 20 + **2**, that is to say 26*

(Or to 22 we add the **4** from 24 = 20 + **4** that is to say 26).
We multiply this 26 result by the close number, which is 20:
26 x 20 = 520

We multiply each other the differences to 20 from 24 and 22, which is 4 x 2 = 8.
We add this result to the previous result:
520 + 8 = 528.

We conclude that 24 x 22 = 528

***

Calculate 56 x 52

We notice that these 2 numbers are close to 50, in fact:
56 = 50 + 6
52 = 50 + 2

To 56 we add the **2** from 52 = 50 + **2**, that is to say: 58.
(Or to 52 we add the **6** from 56 = 50 + **6**, that is to say: 58)
We multiply this result, 58, by the closed number, which is 50:

We saw that in order to multiply by 50, it was easier to multiply by 100 and then divide by 2:

58 x 100 = 5800 and 5800 / 2 = 2900.
So 58 x 50 = 2900

We multiply by each other the differences to 50 from 56 and 52, which is 6 x 2 = 12.
We add this result to the previous result: 2900 + 12 = 2912

We conclude that 56 x 52 = 2912

**

We can extend the method to larger numbers; let's calculate 232 x 211.

We notice that these 2 numbers are close to 200, in fact:
232 = 200 + 32
211 = 200 + 11

To 232 we add the **11** from 211 = 200 + **11**, that is to say: 243
(Or to 211 we add the **32** from 232 = 200 + **32**, that is to say: 243)
We multiply this 243 result by the closest number, which is 200:

We saw that in order to multiply by 200, it was simpler to multiply by 100 and then by 2:

243 x 100 = 24300 and 24300 x 2 = 48600. So, 243 x 200 = 48600.
We multiply by each other the differences to 200 from 232 and 211, which is 32 x 11 = 352 (with the multiplication by 11 method studied before).
We add this result to the previous result: 48600 + 352 = 48952

We conclude that 232 x 211 = 48952.

Exercises:

Solve these multiplications by using the close number method:

a/ 43 x 45
b/ 66 x 64
c/ 332 x 306

Answers:
a/ 1935
b/ 4224
c/ 101592

The multiplication by close number method is possible when we multiply between them

the same larger two numbers having the same first digit.

Multiply by decomposing: simplify calculation

When we need to solve a multiplication, the first reflex to adopt is to look for a way to simplify the calculation. We previously saw very powerful methods that simply allow solving complex calculations. This way, multiplying by 2, by 4, by 8, by 5, by 25, by 50 and even by 11 can be done simpler than all other type of calculation.

That is why the decomposing method is going to be attached to the given operation modification in order to make digits appear from which we have a trick to multiply easier.

Illustration:

Face to a given multiplication, the first stage consists in defining which number(s) we are going to create in order to simplify the calculation:

We can try to create a 2, a 4, an 8, a 5, a 25, a 50, an 11, it means, two close numbers.

Calculate for example 32 x 22

We can notice here that 22 = 11 x 2, we can then create two close numbers that are 11 and 2!

So, 32 x 22 = 32 x 11 x 2.
By the previous studied method, 32 x 11 = 352,
And 352 x 2 = 704,
So, 32 x 22 = 704.

***

Calculate 16 x 4.5

We can notice that 16 = 8 x 2.
So, 16 x 4.5 = 8 x 2 x 4.5 = 8 x 9 = 72.

***

Calculate 13 x 150

We can notice here that 150 = 50 x 3. We can then create a friend number, which is 50!

So, 13 x 150 = 13 x 3 x 50 = 39 x 50.
By the previous studied method, multiplying by 50 would be the same as multiplying by 100 and then dividing by 2:
39 x 100 = 3900
and 3900 / 2 = 1950.

So, 13 x 150 = 1950.

***

Calculate 66 x 32

We can notice here that 66 = 33 x 2. We can then create a friend number, which is 2, but mostly we have two close numbers, which are 33 and 32.

So, 66 x 32 = 2 x 33 x 32
By the close number method, we can calculate 33 x 32:

We notice that these 2 numbers are close to 30, in fact:
33 = 30 + 3
32 = 30 + 2

To 33, we add the **2** from 32 = 30 + **2**, that is to say: 35.
(Or to 32, we add the **3** from 33 = 30 + **3**, which is to say 35)
We multiply this 35 result by the close number, which is 30:
35 x 30 = 1050

We multiply between them the last digits from 33 and 32, that is 3 x 2 = 6.
We add this result to the previous result:
1050 + 6 = 1056.

We conclude that 33 x 32 = 1056.

So, 66 x 32 = 2 x 1056 = 2112.

Exercises:

Solve these multiplications by decomposing numbers:

a/ 66 x 15
b/ 126 x 61
c/ 75 x 52

Answers:
a/ 66x15 = 11x6x15 = 11x90 (the multiplication by 11 method) → 990
b/ 126x61 = 2x63x61 (the closed number method) → = 7686
c/ 75x52 = 3x25x52 (the multiplication by 25 method) → 3900

In order to simply a multiplication, try to create numbers that allow applying a calculation method:
Multiplication by 2, 4 or 8,
Multiplication by 5, 25 or 50,
Multiplication by 11,
Multiplication by close numbers.

Simplify digits to solve a multiplication

When you do not get to apply the previous technique consisting of creating numbers in the calculation, numbers for which you have a calculation trick, it could be life-saving to try the simplification method.

Illustration:

Let's try to mentally calculate 53 x 89.

It is difficult in this example to decompose the calculation in order to create a 2, 4, 8, 11, 5, 25 or 50.
It is difficult too to create close numbers.

On the contrary, we can note that 89 = 90 – 1 and it is simpler to mentally multiply a number by 90 than by 89.

So, 53 x 89 = 53 x (90 – 1) = 53 x 90 – 53 x 1 = 4770 – 53 = 4717.

***

In the same way, we can calculate 41 x 17.

Noting that 41 = 40 + 1, we can write that 41 x 17 = (40 + 1) x 17 = 40 x 17 + 1 x 17 = 680 + 17 = 697.

Exercises:

Solve these multiplications by simplifying numbers:

a/ 62 x 35
b/ 48 x 15
c/ 23 x 34

Answers:
a/ 62x35=(60+2)x35=60x35+2x35=2100+70=2170
b/ 48x15=(50-2)x15=50x15-2x15=750-30=720
c/ 23x34=(20+3)x34=20x34+3x34=680+102=782

In order to simplify a multiplication, if the decomposing method fails, we can try to apply the simplification method.

Cut out numbers in order to solve a multiplication

As we previously saw in additions and subtractions, cutting out a multiplication number can be a very efficient method in order to simplify calculations.

This technique is particularly adapted when we multiply a larger number by a short number.

Illustration:

Try to mentally calculate 12331528 x 3.

This type of calculation can be solved by multiplying from left to right, the method that we previously studied. For all that, some multiplication stages in this operation are going to create remainders (Ex. 5 x 3 then 8 x 3) that are going to make the mental calculation more delicate in a larger number.

The recommended method here is to cut out the larger number in shorter ones of only one or two digits. The cutting out will be judicious in order to limit to the maximum multiplications creating remainders.

In our example, we can in this way solve the next cutting out:

12331528 x 3 = 12//33//15//28 x 3 and to solve these multiplications:
12 x 3 = 36
33 x 3 = 99
15 x 3 = 45
28 x 3 = 84

Putting results end-to-end, we get 12331528 x 3 = 36994584.

***

In the same way, we can calculate 211523 x 4.

Applying the cutting out 21//15//23 x 4 = 84//60//92 from which we conclude that 211523 x 4 = 846092.

Exercises:

Solve these multiplications by cutting out numbers:

a/ 19241 x 4
b/ 181217 x 5
c/ 342514 x 3

Answers:
a/ 19241 x 4 = 19//24//1 x 4 = 76//96//4 = 76964
b/ 181217 x 5 = 18//12//17 x 5 = 90//60//85 = 906085
c/ 342514 x 3 = 34//25//14 x 3 = 102//75//42 = 1027542

In order to solve a multiplication implying a large number and a short number, we can wisely cut out the larger one in order to simplify the calculation.

Mental multiplication of very large numbers: it is possible

The previous cutting out method works very well when there is a large number by a short number multiplication, frequently limited to a digit. This tells us that, after having surprised your friends with one of these kinds of calculations, these ones will challenge you to a more complex mental solution.

How will you react when the teasing of them will cause them to ask you to mentally solve this calculation: 121423 x 100002?

Do not worry, applying step by step the method I'm going to teach you, and with a lot of practice, you will soon get to make the achievement of mentally getting this calculation's result!

Illustration:

In order to illustrate the method, we are going to star from a more modest example.

Let's try to mentally give the result of 768 x 997.

Using the close number method in order to say that:
768 is closed to 1000 because 768 = 1000 – 232.
997 is closed to 1000 because 997 = 1000 – 3.

From 768 we subtract the **3** from 997 = 1000 – **3**, that is to say: 765.
(Or to 997 we subtract the **232** from 768 = 1000 - **232** that is to say too 765 even it is more complicated to calculate).
We multiply this 765 result by the close number, which is 1000:
765 x 1000 = 765000

We multiply each of the differences to 1000 from 768 and 997, which is 232 x 3 = 696.
We add this result to the previous result:
765000 + 696 = 765696.

We conclude that 768 x 997 = 765696.

Note that it is enough to join the first found result, 765, to the second found result, 696, in order to get the final result 765696.

***

In the same way, we can calculate 121423 x 100002.

Use the close number method, in order to say that:
121423 is closed to 100000 because 121423 = 100000 + 21423
100002 is closed to 100000 because 100002 = 100000 + 2

To 121423 we add the **2** from 100002 = 100000 + **2** that is to say 121425
(Or to 100002 we add the **21423** from 121423 = 100000 + **21423** that is to say 121425, even if it is more complicated to calculate).
We multiply this 121425 result by the close number, which is 100000:
121425 x 100000 = 12142500000.

We multiply each other by the differences to 100000 from 121423 and 100002, which is 21423 x 2 = 42846.
We add this result to the previous result:
12142500000 + 42846 = 12142542846.

We conclude that 121423 x 100002 = 12142542846.

Note that it is enough to join the first found result, 121425, to the second found result, 42846, in order to get the final result 12142542846.

Exercises:

Solve these multiplications by the large number method:

a/ 1123 x 1002
b/ 886 x 998
c/ 8952 x 9995

Answers:
a/ 1125246
b/ 884228
c/ 89475240

The examples below all have a common point. Calculations that they create imply only numbers having the same digits as the other. We have in this way calculated a multiplication with two numbers of 3 digits and then, a multiplication with two numbers of 6 digits. This implies that the used close number is the same for the calculation of two numbers: this way, for 768 x 997 the close number is 1000, for 121423 x 100002 the close number is 100000. All of that makes it interesting to solve multiplications of the type of 9997 x 96, it means, that it implies two numbers that do not have the same quantity of digits.

The method to solve this kind of multiplication is a little bit more complicated, but it is easy assimilated with a little practice.

Illustration:

Let's try to mentally get the result of 9997 x 96.

We can use the close number method in order to say that:
9997 is closed to 10000 because 9997 = 10000 – 3.
96 is close to 100 because 96 = 100 – 4

Let's mentally show the calculation as follows:

Numbers multiplied between them	Difference related to close number
9997	- 03
96	- 04

*We begin with the larger number (9997) to which we subtract the **4** from 96 = 100 – **4**.*

78

The subtlety consists in verifying that, in the chart, 96 is lined up with the 99 from 9997. This means that the 4 must be subtracted from the 99 from 9997 and not to the 7 from 9997.

We get, then, the 9**5**97 number.

Next, we multiply each of the differences related to close numbers, which are 03 x 04 = 12

We put end-to-end the calculated results; that is to say: 9997 x 96 = 959712.

■■

Let's try to mentally get the result of 10121 x 1003.

Inspiring us on the close number method in order to say that:
10121 is closed to 10000 because 10121 = 10000 + 121.
1003 is closed to 1000 because 10003 = 1000 + 3.

Let's mentally show the calculation as follows:

Numbers multiplied between them	Difference related to the close number
10121	+ 121
1003	+ 003

We begin with the larger number (10121), to which we add the **3** from 1003 = 1000 + **3**.

The subtlety consists in verifying that, in the chart, 1003 is lined up to the 1012 from 10121. This means that the 3 must be added to 1012 from 10121 and not to 10121.

We get then the number 101**5**1.

Next, we multiply the differences related to close numbers to each other, which are 121 x 003 = 363

We put end-to-end the calculated results; that is to say: 10121 x 1003 = 10151363.

In order to solve a multiplication implying two large numbers, we can core to the close number method from 1 000, 10 000 or 100 000.

Immediately estimate a multiplication result

Sometimes, you will only need to find the first digit of a result, and the number of zeros that follow it, and not all important digits of the result.

There is a simple method that will allow you to determine in a twinkling of an eye the magnitude of a multiplication result.

Illustration:

Let's try to mentally give the magnitude of 32 x 51

Mentally, it can be represented as:
32 is close to 30,
51 is close to 50

Our calculation will give a close result to this one: 30 x 50 = 1500.

In fact, 32 x 51 = 1632.

***

Let's try to mentally give a magnitude of 496 x 42

Mentally, it can be represented as:
496 is close to 500,
42 is close to 40

Our calculation will give a close result to this one: 500 x 40 = 20000.

In reality, 496 x 42 = 20832.

Exercises:

Give an estimation of the magnitude of the multiplication results:

a/ 1523 x 197
b/ 691 x 812
c/ 10320 x 298

Answers:
a/ 1500 x 200 = 300000
b/ 700 x 800 = 560000
c/ 10000 x 300 = 3000000

> **Round up the two numbers of a multiplication to the close numbers, allowing to get a magnitude of this operation.**

Verify a multiplication result

As in an addition and a subtraction, it is possible to use the digital root in order to verify a multiplication result.

Illustration:

Previously, we calculated 121423 x 100002 = 12142542846.

Let's calculate the digital roots of the above numbers:
121423 = 1+2+1+4+2+3 = 13 → 1+3 = 4.
100002 = 1+0+0+0+0+2 = 3.
12142542846=1+2+1+4+2+5+4+2+8+4+6 = 39 → 3+9 = 12 → 1+2 = 3.

The operation we did (121423 x 100002) is a multiplication.

We multiply then the digital root of 121423 (which is 4) and the one of 100002 (which is 3); that is to say 12. The digital root of 12 is 1+2 = 3.

We verify that the digital root is the same as the result digital root 12142542846 (which is 3).

If we get the same digital root, the calculation is probably correct; if it is not the case, it is certain that the calculation is wrong.

Exercises:

Determine, by the digital root, if these calculations are wrong:

a/ 125 x 341 = 42625
b/ 97 x 651 = 63147
c/ 1024 x 422 = 432228

Answers:
a/
Determine the digital roots:
125 : 1+2+5=8
341 : 3+4+1=8
426256 : 4+2+6+2+5=19→1+9=10→1+0=1
Multiplication of the digital roots:
8x8=64→6+4=10→1+0=1
The digital roots are the same; calculation is correct.

b/
Determine the digital roots:
97 : 9+7=16→1+6=7
651 : 6+5+1=12→1+2=3
63147 : 6+3+1+4+7=21→2+1=3
Multiplication of the digital roots:
7x3=21→2+1=3
The digital roots are the same; calculation is correct.

c/
Determine the digital roots:
1024 : 1+0+2+4=7
422 : 4+2+2=8
43228 : 4+3+2+2+8=19→1+9=10→1+0=1

Multiplication of the digital roots:
7x8=56→5+6=11→1+1=2
The digital roots are not the same; calculation is certainly wrong.

Determine if a multiplication result is wrong by using the digital roots.

Sum up the multiplication techniques

We saw that the techniques and tricks to mentally solve multiplications are many. At the beginning, you will probably feel yourself lost in choosing the technique that will allow you to easily solve the proposed calculation.

I propose you to look back on the different methods that we have discovered together with its main characteristics:

Multiplication by 2, 4, 8:
This technique allows us to easily multiply a number by 4 and by 8.

Multiplication by 5, 25 and 50:
This technique allows us to easily solve a multiplication in which appears one of these numbers: 5, 25 or 50.

Multiplication by 11:
This technique allows multiplying a number by 11 in the twinkling of an eye.

Multiplication by close numbers:
This technique allows multiplying between them 2 or 3 digit numbers that have the same first digit.

Multiplication by decomposing:
This technique consists in simplifying a multiplication, making appear numbers that allow applying one of the above methods.

Multiplication by simplifying:
In the case of multiplications between of 2-digit numbers, this technique can be applied, in case of difficulty, to use the decomposing method. The simplifying method aims to create in the calculation a number multiple of 10 (10, 20, 30, 40 ...., 90) with which it is easier to mentally solve a multiplication.

Multiplication by cutting out:
This technique is particularly adapted to multiplications between a large number and a one digit number. It aims to cut out the large number by packages of shorter numbers that are easier to multiply.

Multiplication of very large numbers:
This technique is very efficient in order to solve multiplications implying large numbers. It is based on the close number method of 1000, 10000, and 100000.

For the purpose of helping you to apply these techniques, I propose to you, here below, some examples that will allow you to understand the approach to be used in order to determine which method is the most efficient technique to solve the proposed calculation. Practicing regularly with the concrete cases, this process will become automatic, and you will very soon be capable of selecting the more pertinent method, in an intuitive way, by observing the numbers that compose the calculation.

Illustration:

Calculate 532 x 3

3 is not a number with which there is a particular multiplication trick → elimination of astute methods.
532 can't be easily decomposed in order to create a friend number or close numbers → elimination of the decomposing method.
532 x 3 is not a large number multiplication → elimination of the close number methods.
532 x 3 is a large number multiplied by a one digit number → utilization of the cutting out method.

We can write 532 x 3 = 5//3//2 x 3 = 5x3 // 3x3 // 2x3
So, 532 x 3 = 15 // 9 // 6 which is 1596.

***

Calculate 63 x 4

4 is a number with which there is a particular multiplication trick → utilization of the multiplication by 4 method.

Multiplying by 4 would be the same as multiplying two times by 2.

So, 63 x 2 = 126 and 126 x 2 = 252
Therefore, 63 x 4 = 252.

***

Calculate 75 x 6

6 is not a number with which there is a particular multiplication trick → elimination of the astute methods.
75 can't be easily decomposed in order to make appear friend numbers or close numbers → utilization of the decomposing method.

It is enough to notice that 75 = 25 x 3.

So, 75 x 6 = 25 x 3 x 6 = 25 x 18.

In order to multiply by 25, we have learned that it is enough to multiply by 100 and then divide two times by 2; that is to say:

18 x 100 = 1800
1800 / 2 = 900
900 / 2 = 450

So, 75 x 6 = 450.

***

Calculate 126 x 62

Both numbers are numbers with which there is a particular multiplication trick → elimination of the astute methods.
126 can be easily decomposed in order to create friend numbers or close numbers → utilization of the decomposing method.

It is enough to notice that 126 = 2 x 63.

So, 126 x 62 = 2 x 63 x 62.

We observe that 63 and 62 are close numbers having the same first digit.

We can then use the close number method in order to calculate 63 x 62:

We notice that these 2 numbers are close to 60, in fact:
63 = 60 + 3
62 = 60 + 2

To 63 we add the **2** from 62 = 60 + **2**; that is to say: 65.
(Or to 62 we add the **3** from 63 = 60 + **3**; that is to say, too: 65)
We multiply this 65 result by the close number, which is 60:
65 x 60 = 3900

We multiply between them the last digits from 63 and 62 which are 3 x 2 = 6.
We add this result to the previous result:
3900 + 6 = 3906.

We conclude that 63 x 62 = 3906.

So, 126 x 62 = 2 x 3906 = 7812.

***

Calculate 123 x 104

Both numbers are numbers with which there is a particular multiplication trick → elimination of the astute methods.
None of the numbers can be decomposed in order to create friend numbers or close numbers → elimination of the decomposing methods.
123 x 104 is a large number multiplication → utilization of the close number method.

123 is close to 100 because 123 = 100 + 23
104 is close to 100 because 104 = 100 + 4

To 123 we add the **4** from 104 = 100 + **4;** that is to say: 127.
(Or to 104 we add the **23** from 123 = 100 + **23;** that is to say, too: 127)
We multiply this 127 result by the close number, which is 100:
127 x 100 = 12700.

We multiply each by the differences to 100 from 123 and 104 which is 23 x 4 = 92.
We add this result to the previous result:
12700 + 92 = 12792.

We conclude that 123 x 104 = 12792.

Note that it is enough to join the first found result, 127, to the second one, 92, in order to get the final result 12792.

***

Calculate 998 x 97

Both numbers are numbers with which there is a particular multiplication trick → elimination of the astute methods.

None of them can be easily decomposed in order to create friend numbers or close numbers → elimination of the decomposing methods.

998 x 97 is a large number multiplication → utilization of the close number method.

998 is close to 1000 because 998 = 1000 – 2.
97 is closed to 100 because 97 = 100 – 3.

Mentally we show the calculation as follow:

Numbers multiplied between them	Difference related to the close number
998	- 02
97	- 03

*We start with the larger number (998) to which we subtract the **3** from 97 = 100 – **3**.*

*The subtlety is in observing that, in the chart, 97 is line up with the 99 from 998. This means that the **3** should be* subtracted from 99 from 998 and not to 998.

We get then the number 9**6**8.

Next, we multiply each by the differences related to close numbers, which are 02 x 03 = 06

We put the results end-to-end; that is to say: 998 x 97 = 96806.

Exercises:

Determine the method to be used and then solve these calculations:

a/ 226 x 50
b/ 21 x 534
c/ 151 x 55
d/ 42 x 42
e/ 1121 x 1005

Answers:
a/ multiplication by 50 → 11300
b/ simplification: 21 x 534 = 20 x 534 + 1 x 534 = 11214
c/ decomposing (55 = 11 x 5) then multiplication by 11 and by 5 → 8305
d/ multiplication by close numbers having the same first digit → 1764
e/ multiplication of very large numbers → 1126605

Division techniques

Division is an operation that will be useful to you in all situations where you will need to share. This operation will be of interest to the child who collects Pokémon cards and who wants to share 26 cards among 4 friends, but also to the mother who will share 75 bonbons among her three kids. According to the quantities to be shared, calculation will be more or less easy. The purpose of this chapter is to allow you to mentally solve most of the divisions you will find in your daily life, but also, if you want to, to solve complex divisions in the twinkling of an eye in order to impress your entourage.

This number can be divided? Divisibility criteria

In my introduction to division, I mentioned two examples in which I'm going to briefly stop.

The first case I mentioned is that one about the kid who collects Pokémon cards. He has 26 doubles that he wants to share with his 4 friends. The question is to determine how many cards he will give to each friend. With 26 cards, he can make four 6-card packages. That is 24 cards, and he will keep 2 cards.

The second situation is the one of the mom who wants to share 75 bonbons between her 3 children. We talk here about determining how many bonbons she can distribute to each kid. The answer is given by the division result of 75 by 3, which is 25. She can then give 25 bonbons to each kid, and she will keep none.

In the first case, we divided 26 by 4, and it left 2.
In the second case, we divided 75 by 3 and it left nothing.

We will say this way that 26 is not divisible by 4 because when we solve the calculation there is one remainder. On the contrary, we will say that

75 is divisible by 3 because when we solve the calculation there are not remainders.

When we want to mentally solve a division, it is very useful to know, before we even start the calculation, if there will be remainders in this division or not. There are very simple methods in order to determine if a number is divisible by 2, 3, 4 ... until 13. This method is called the **divisibility criteria**.

Divisibility criteria	Example
Divisibility by 2: A number is divisible by 2 if its last digit is paired (it finishes by 0, 2, 4, 6 or 8)	126 is divisible by 2. 133 is not divisible by 2.
Divisibility by 3: A number is divisible by 3 if the addition of its digits is a multiple of 3.	131 is not divisible by 3 because 1+3+1=5, and 5 is not a multiple of 3. 531 is divisible by 3 because 5+3+1=9, and 9 is a multiple of 3.
Divisibility by 4: A number is divisible by 4 if its two last digits are a multiple of 4.	311 is not divisible by 4 because 11 is not a multiple of 4. 624 is divisible by 4 because 24 is a multiple of 4.
Divisibility by 5: A number is divisible by 5 if its last digit is a 0 or a 5.	234 is not divisible by 5 because its last digit is 4.

	990 is divisible by 5 because its last digit is 0.
Divisibility by 6: A number is divisible by 6 if it is at the same time divisible by 2 and by 3.	741 is not divisible by 6 because it is not divisible by 2 (but it is divisible by 3). 234 is divisible by 6 because it is at the same time divisible by 2 and by 3.
Divisibility by 7: A number is divisible by 7 if the difference between the number of tens and the double of the unit digit is divisible by 7.	176 is not divisible by 7 because $17 - 2 \times 6 = 17 - 12 = 5$ is not divisible by 7. 553 is divisible by 7 because $55 - 2 \times 3 = 55 - 6 = 49$ is divisible by 7.
Divisibility by 8: A number is divisible by 8 if ht + (u/2) is divisible by 4. In the number: h is the digit of hundreds, t the digit of tens and u the digit of units.	834 is not divisible by 8 because ht + (u/2) = 83 + (4/2) = 85 is not divisible by 4. 616 is divisible by 8 because ht + (u/2) = 61 + (6/2) = 64 is divisible by 4.
Divisibility by 9: A number is divisible by 9 if its digit addition is a multiple of 9.	445 is not divisible by 9 because 4+4+5=13 and 13 is not divisible by 9. 756 is divisible by 9 because 7+5+6=18 and 18 is divisible by 9.

Divisibility by 10: A number is divisible by 10 if its last digit is 0.	849 is not divisible by 10 because its last digit is 9. 320 is divisible by 10 because its last digit is 0.
Divisibility by 11: A number is divisible by 11 if the difference between the pair digit addition and the impair digit addition is divisible by 11.	13<u>5</u>4 is not divisible by 11 because <u>1</u>+<u>5</u>=6 and 3+4=7 and 7-6=1 is not divisible by 11. 13<u>6</u>4 is divisible by 11 because <u>1</u>+<u>6</u>=7 and 3+4=7 and 7-7=0 is divisible by 11.
Divisibility by 12: A number is divisible by 12 if it is at the same time divisible by 3 and by 4.	525 is not divisible by 12 because it is not divisible by 4 (but it is divisible by 3). 156 is divisible by 12 because it is at the same time divisible by 3 and by 4.
Divisibility by 13: A number is divisible by 13 if the addition of the number of tens and the quadruple of the unit digits is divisible by 13.	426 is not divisible by 13 because 42 + 4x6 = 42+24=68 is not divisible by 13. 637 is divisible by 13 because 63 + 4x7 = 63+28=91 is divisible by 13. 91 is divisible by 13 because 9 + 4x1 = 9+4 = 13 is divisible by 13.

Criteria given in the above chart are relatively simple to apply, and they give a quick answer when it is about determining if a number is divisible by another number between 1 and 13. Nevertheless, what about the divisibility of 4913 by 17 or of 3141 by 59?

It is again sufficiently easy to adopt an answer to this question thanks to the method called the **zero method**:

Illustration:

We need to determine if 4913 is divisible by 17.
The method consists in adding or subtracting numbers that are multiples of 17 to the number 4913 in order to make zeros appear.

For example: 4913 + 17 = 4930.

We subtract then the 0; that is to say 493.

We repeat then the first stage: 493 + 17 = 510. Then we subtract the 0, and it remains then 51.

We notice that 51 = 3 x 17.

So, 4913 is divisible by 17.

Determine if 3141 is divisible by 59.
3141 + 59 = 3200. By subtracting the 0, it remains 32.

32 is not divisible by 59; we conclude that 3141 is not divisible by 59.

Exercises:

Determine by the zero method if these numbers are divisible:

a/ 22222 by 41
b/ 2777 by 23

Answers:
a/ 22222 − 2x41 = 2222 − 82 = 22140 → 2214 − 4x41 = 2214 − 164 = 2050 → 205 − 5x41 = 205 − 205 = 0 So, 22222 is divisible by 41.
b/ 2777 + 23 = 2800 → 28 is not divisible by 23, so 2777 is not divisible by 23.

Divide by 2, 4, and 8 before the calculator

Since division is the reciprocal of the multiplication, the mathematic tricks with the numbers 2, 4, and 8 that we studied in the setting of multiplication can be evidently transposed to division, in this way:

Dividing by 4 would be the same as dividing two times by 2.
Dividing by 8 would be the same as dividing three times by 2.

In this way, 64 / 4 = 64 / 2 / 2 = 32 / 2 = 16

And 152 / 8 = 152 / 2 / 2 / 2 = 76 / 2 / 2 = 38 / 2 = 19.

Exercises:

Solve these divisions:

a/ 92 / 4
b/ 184 / 8
c/ 1296 / 4

Answers:
a/ 23
b/ 23
c/ 324

> *Dividing by 2 is an easy to solve operation.*
> *Dividing by 4 would be the same as dividing two times by 2,*
> *Dividing by 8 would be the same as dividing three times by 2.*

Dividing by 5, 25, and 50 before the calculator

When studying the multiplication techniques, we saw that:

In order to multiply by 5, you begin by multiplying by 10 and then dividing by 2.
In order to multiply by 50, you begin by multiplying by 100 and then dividing by 2.
In order to multiply by 25, you begin by multiplying by 100 and then dividing two times by 2.

Since division is the reciprocal of the multiplication, it is easy to keep the way of dividing by 5, by 50 and by 25; in fact:

In order to divide by 5, you begin by multiplying by 2 and then dividing by 10.
In order to divide by 50, you begin by multiplying by 2 and then dividing by 100.
In order to divide by 25, you begin by multiplying two times by 2 and then dividing by 100.

Illustration:

We need to calculate 2250 / 5

Let's begin by calculating 2250 x 2 = 4500.
Then, we calculate 4500 / 10 = 450.
So, 2250 / 5 = 450

Calculate 7250 / 50
Let's begin by calculating 7250 x 2 = 14500.
Then, we calculate 14500 / 100 = 145.

Calculate 9650 / 25
Let's begin by calculating 9650 x 2 = 19300.
Then, we calculate 19300 x 2 = 38600.
And then again: 38600 / 100 = 386.

Exercises:

Solve these divisions:

a/ 6250 / 5
b/ 11300 / 50
c/ 8650 / 25

Answers:
a/ 1250
b/ 226
c/ 346

Dividing by 5 would be the same as multiplying by 2 and then dividing by 10.
Dividing by 50 would be the same as multiplying by 2 and then dividing by 100.
Dividing by 25 would be the same as multiplying two times by 2 and then dividing by 100.

Divide by 9 before the calculator

In the multiplication chapter, I made you love number 11, showing you how simple it is to multiply a number by 11. This time, we are going to do the same for the division but with number 9.

Illustration:

Calculate 20403 / 9

We put down the first digit starting from the left, which is 2.
We add this digit (2), to the digit on its right (0); that is 2+0=2.
We add this digit (2) to the next digit on its right (4); that is 2+4=6.
We add this digit (6) to the next digit on its right (0); that is 6+0=6.
We add this digit (6) to the next digit on its right (3); that is 6+3=9.

The first 4 digits give the result. The last digit gives the remainder; that is 2266, remainder 9. It means 2267, remainder 0.

We conclude then that 20403 / 9 = 2267.

**

This also works with a larger number.
Calculate 124523 / 9

We put down the first digit starting from the left, which is 1.
We add this digit (1), to the digit on its right (2); that is 1+2=3.
We add this digit (3) to the next digit on its right (4); that is 3+4=7.
We add this digit (7) to the next digit on its right (5); that is 7+5=12.
We add this digit (12) to the next digit on its right (2); that is 12+2=14.
We add this digit (14) to the next digit on its right (3); that is 14+3=17.

The first 5 digits give the result; the last digit gives the remainder, that is

1 3 7 (12) (14) remainder (17).

When a number has 2 digits, the first one of these digits is added to the previous number, so we have:

1
3

7 + ***1*** *(1 comes from **1**2 following 7), which is 8.*
2 + ***1*** *(1 comes from **1**4 following 12), which is 3.*
4

Remainder 17

We conclude then that 124523 / 9 = 13834 remainder 17 or 13835 remainder 8.

Exercises:

Solve these divisions:

a/ 693 / 9
b/ 3170 / 9
c/ 2205 / 9

Answers:
a/ 77
b/ 352 remainder 2
c/ 245

I agree that, according to the size of the number to be divided by 9, this technique is not necessarily the simplest to be mentally applied. Nevertheless, with a little bit of practice, you should be capable of solving this kind of division with a paper and a pencil more quickly than by the classic division method as the one learned at school.

Decompose numbers in order to solve a division

As we studied for the multiplication, the cutting out technique for a number can be transposed to division.

This technique is particularly adapted when we divide a number by a short number between 1 and 9 in order to use common multiplication tables.

Illustration:

Let's try to mentally calculate 4249 / 7.

The recommended method here is to cut out the larger number in shorter numbers of only 1 or 2 digits. The cutting out should be wise in order to make it simple to solve division.

In our example, we can do this in this way to solve the next cutting out:

4249 / 7 = 42//49 / 7, and solving these divisions:
42 / 7 = 6
49 / 7 = 7

By putting these results end-to-end, we get 4249 / 7 = 67.

***

In the same way, we can calculate 24456 / 3.

Solving the cutting out 24//45//6 / 3 = 08//15//2, from where we conclude that 24456 / 3 = 8152.

Exercises:

Solve these divisions by cutting out numbers:

a/ 102545 / 5
b/ 1218144 / 2
c/ 48365412 / 6

Answers:
a/ 102545 / 5 = 10//25//45 / 5 = 2//05//08 = 20508
b/ 1218144 / 2 = 12//18//14//4 / 2 = 06//09//07//2 = 609072
c/ 48365412 / 6 = 48//36//54//12 / 6 = 08//06//09//02 = 8060902

In order to solve a division implying a large number and short number, we can wisely cut out the larger number in order to simplify the calculation.

Divide by means of close numbers

The next technique allows very easily solving divisions by implying close numbers of 100, 1000, or 10000.

Illustration:

Let's try to calculate 25485 / 8676.

It is a question of notice that 8676 is a close number of 10000.
Using the technique studied in the previous chapter, "Subtract numbers to 10, 100, 1000 in the twinkling of an eye", we can easily determine that 10000 − 8676 = 1324.

We will show these different numbers in the next way:

B = Base (Y zeros)		
A = Divisor	E = Other digits of the number to be divided	D = Y digits of the number to be divided
C = (Base − Divisor)		F = C X E
		G = D + F

Chart explanation:

A: We write here our operation divisor, in our example, it is 8676.

B: We write here the base to which the divisor is close to. This will be a number like 100, 1000, 10000 etc.... In this case, it will be 10000. Let's notice that 10000 has 4 zeros.

C: We write here the subtraction result of the base and the divisor. In this case, 10000 − 8676 = 1324.

D: We write here the digits of the number to be divided. In this box, there are as many digits as zeros in the base. In our example, the base is 10000; there are then 4 zeros. Therefore, we write in this box, the 4 last digits of the number to be divided (25485), which is 5485.

E: We write here all the other digits of the number to be divided that are not written in D. In this case, it remains the digit 2.

F: We write here the multiplication result between the number written in C and the number written in B. In this case, we multiply 1324 x 2 = 2648.

G: we write the addition result of the number written in F and the number written in D. In this case, we add 2648 + 5485 = 8133.

We finally get the next chart:

10 000			
8 676		2	5485
1 324			2648
			8133

The solved division result is given in the E box and the remainder digit in the G box.

We conclude that 25485 / 8676 = 2 remainder 8133.

***

In the same way, we can calculate 22321 / 7999.

Filling the chart as above explained, we get:

10 000			
7 999		2	2321
2 001			4002
			6323

We conclude that 22321 / 7999 = 2 remainder 6323.

Exercises:

Solve these divisions according to the same method:

a/ 125654 / 98751
b/ 497856 / 95454
c/ 95621 / 8753

Answers:
a/ 1 remains 26903
b/ 5 remains 20586
c/ 10 remains 8091

The close number method allows simplifying a division because the divisor is close to numbers like 100, 1000, 10000 etc....

Verify a division result

As with the other operations, it is possible to use the digital root in order to verify a division result.

Illustration:

Previously, we calculated 25485 / 8676 = 2 remainder 8133.

That would be the same as writing 8676 x 2 + 8133 = 25485.

Let's calculate the above digital numbers:
8676 = 8+6+7+6 = 27 → 2+7 = 9
8133 = 8+1+3+3 = 15 → 1+5 = 6
25485 = 2+5+4+8+5 = 24 → 2+4 = 6

In the operation 8676 x 2 + 8133, we replace each number by its digital root, which is to say 9 x 2 + 6 = 18 + 6 = 24 → 2 + 4 = 6

We verify that the digital root is the same as the digital root result 25485 (which is 6).

If we get the same digital root, the calculation is probably correct; if it is not the case, the calculation is certainly wrong.

Exercises:

Determine, by the digital root, if these calculations are wrong:

a/ 12453 / 256 = 48 reste 165
b/ 9852 / 128 = 76 reste 124
c/ 5644 / 344 = 16 reste 135

Answers:
a/
Determining digital roots:
48: 4+8=12 → 1+2=3
256: 2+5+6=13 → 1+3=4
165: 1+6+5=12→1+2=3
12453: 1+2+4+5+3=15 → 1+5=6
Calculate digital roots:
256 x 48 + 165: 4 x 3 + 3 = 15 → 1+5=6
Digital roots are the same, calculation is correct.
b/
Determining digital roots:
76: 7+6=13→1+3=4
128: 1+2+8=11→1+1=2
124: 1+2+4=7
9852: 9+8+5+2=24 →2+4=6
Calculate digital roots:
128 x 76 + 124 : 2 x 4 + 7 = 15 → 1+5=6
Digital roots are the same, calculation is correct.
c/
Determining digital roots:
16: 1+6=7
344: 3+4+4=11 → 1+1=2
135: 1+3+5=9
5644: 5+6+4+4=19 → 1+9=10 → 1+0=1
Calculate digital roots:
344 x 16 + 135: 2 x 7 + 9 = 23 → 2+3=5
Digital roots are not the same, calculation is certainly wrong.

Determine if a division result is wrong by using the digital roots.

Conclusion

We arrive at the end of our trip to the number and mathematic country. I hope you enjoy to go through these pages in my accompanying and that the different techniques we studied have modified your perception of the mental calculation and maybe, to reconcile you to digits.

Albert Einstein said "no problem can be solved from the same level of consciousness that created it". In this way, we saw that the techniques shown in this book allow sufficiently easy to transform an unfriendly operation in a very simpler operation. Look at numbers, transform them, and play with them until they find a form that fits you.

I encourage you to go through these pages in order to assimilate the different methods and especially, to take advantage of any occasion that the daily life gives you to practice. You will realize that very soon mental calculation becomes a second nature to you and soon you will naturally use all these methods without even think about it.

www.ingramcontent.com/pod-product-compliance
Lightning Source LLC
Chambersburg PA
CBHW051547170526
45165CB00002B/909